How to Raise Chickens

A Beginner's Guide to Raising Chickens for Eggs in Your Backyard or Homestead

Text Copyright © Lightbulb Publishing

All rights reserved. No part of this guide may be reproduced in any form without permission in writing from the publisher except in the case of brief quotations embodied in critical articles or reviews.

Legal & Disclaimer

The information contained in this book and its contents is not designed to replace or take the place of any form of medical or professional advice; and is not meant to replace the need for independent medical, financial, legal or other professional advice or services, as may be required. The content and information in this book has been provided for educational and entertainment purposes only.

The content and information contained in this book has been compiled from sources deemed reliable, and it is accurate to the best of the Author's knowledge, information, and belief. However, the Author cannot guarantee its accuracy and validity and cannot be held liable for any errors and/or omissions. Further, changes are periodically made to this book as and when needed. Where appropriate and/or necessary, you must consult a professional (including but not limited to your doctor, attorney, financial advisor or such other professional advisor) before using any of the suggested remedies, techniques, or information in this book.

Upon using the contents and information contained in this book, you agree to hold harmless the Author from and against any damages, costs, and expenses, including any legal fees potentially resulting from the application of any of the information provided by this book. This disclaimer applies to any loss, damages or injury caused by the use and application, whether directly or indirectly, of any advice or information presented, whether for breach of contract, tort, negligence, personal injury, criminal intent, or under any other cause of action.

You agree to accept all risks of using the information presented in this book.

You agree that by continuing to read this book, where appropriate and/or necessary, you shall consult a professional (including but not limited to your doctor, attorney, or financial advisor or such other advisor as needed) before using any of the suggested remedies, techniques, or information in this book.

Table of Contents

Introduction 1

Chapter One: Why Chickens? 3

Chapter Two: Ready…Get Set… 7

Chapter Three: Ready…Set…and now Go! 15

Chapter Four: What Chickens Eat 19

Chapter Five: Chicken Care 101 23

Chapter Six: Which Breed is Best for You? 39

Chapter Seven: Pecking Order…It's a Real Thing 49

Chapter Eight: The Purpose of Your Flock 53

Chapter Nine: Questions and Answers 65

Closing Comments 71

Introduction

Let's play a little old-fashioned word association game. I say chicken, you say _____. I say eggs, you say _____. I say fried, you say _____.

I'm willing to bet at least two of your three answers were food-related. That's because that's pretty much the only function we associate with chickens. Their purpose is food—in one form or another.

I'm also willing to bet you don't give chickens credit for being anything more than a 'dumb animal.' After all, that saying about chickens not having enough sense to come in out of the rain didn't come from someone who woke up one morning and decided to give these birds a bad name…right?

Wrong!

A little rain never hurt anyone—especially a chicken. The feathers of a chicken are a bit like an oil cloth; they are somewhat waterproof. Even more important, chickens use their 'raincoats' to their advantage on misty, showery days to forage without having to fear predators that don't like the rain. Some people also say the cloudy skies make it harder for predators such as hawks and coyotes to see the chickens.

I know you bought this book for the basic knowledge and information you need to successfully raise chickens for your dining pleasure and possibly as a source of extra income, and I assure you that is what you'll get from it. But I also think it's good (if not just fun) to know a little bit about the animal itself. Knowing your 'chicken facts' will make raising your chickens more enjoyable and will give you something to talk about at farmer's markets or wherever else you go to sell eggs or butcher chickens.

So, before we get into the necessities of successful chicken-raising, take a minute to learn a few fun and unusual facts about your new feathered friends:

- Hens 'talk' to their eggs before they are hatched, teaching them different calls and sounds.
- Chickens are not colorblind. They recognize every color in the rainbow.
- Chickens recognize each other.
- Chickens have a social order of
- Importance and dominance. This is where the phrase 'pecking order' comes from.
- Hens are very protective of their young.
- Chickens cannot fly more than a few seconds at a time and cannot fly any higher than a semi-tall tree.
- There are more chickens than any other type of bird in the world.
- Hens prefer roosters with brightly colored and large combs, and those that strut often and proudly.

How many of these facts did you already know? Don't worry if the answer is 'none' because now you *do* know, and you are about to learn a whole lot more.

https://www.peta.org/action/10-surprising-facts-chickens/
https://www.smithsonianmag.com/science-nature/14-fun-facts-about-chickens-65848556/

Chapter One
Why Chickens?

QUESTION: Why raise chickens?

ANSWER: Why not? Chickens are an excellent introductory species of livestock for novices in the world of agriculture for many reasons. Let's count…

1. Chickens can be raised just about anywhere—even in the city limits of most cities and towns. The term 'backyard chickens' is a literal one. Most city municipalities have ordinances that allow a limited number of backyard chickens. Six is typically the 'magic number' of allowed city chickens, but this is not always true, so **learn your local laws and regulations before purchasing chickens.**

   NOTE: Roosters are usually prohibited within any city limits, and Home Owner Association Covenants and Restrictions may usurp city regulations.

2. Chickens don't require a lot of space.

3. Chickens are fairly low-maintenance. Once you have everything in place, daily feeding, watering, and egg gathering is all that is necessary. (Weekly coop and pen maintenance is also suggested, depending on how your chickens are housed and how much space they have.)

4. Chickens are not expensive to keep and maintain. The investment is relatively small compared to other species of livestock and agricultural hobbies.

5. Chickens can be a good starter project for young children.

    NOTE: Children should not be left unattended with mature roosters or some aggressive breeds.

6. Chickens are a multi-purpose animal. Because they provide eggs and meat, they are a good choice for developing self-sufficiency.

7. Chickens give you a quick return on your investment. It only takes a few months (depending on the breed) for a chick to become an egg-producing hen.

8. Chickens can easily become a source of added income for you and your family. A dozen laying hens can produce a dozen eggs a day. I don't know any family that consumes seven dozen eggs a week, so it is possible to have four to six dozen eggs to sell a week. No, that's not enough to quit your day job, but as you will read later, selling eggs isn't the only way to add income to your budget using your chickens.

9. Chickens help with pest control and waste disposal. They eat ticks and chiggers, as well as several other kinds of small insects. Don't worry, it doesn't affect their health. They also eat watermelon and cantaloupe rinds, lettuce that is wilted or brown, stale cereal (not the sugary kind), and several other types of fruits, veggies, whole grains, and even small bits of meat you would otherwise throw out.

Why Chickens?

Chickens? Why not?

Chapter Two
Ready…Get Set…

The best way to do anything is to do it the best way possible. In other words, don't get your chickens before you are properly prepared. Remember: Anything worth doing is worth doing right.

If you'll remember from the previous chapter, one of the great things about chickens is that they are fairly low-maintenance and start-up costs are low in comparison to other species of livestock. But low-maintenance does NOT mean *no*-maintenance, and small investment does NOT mean zero investment or preparation.

The first thing you need to do is decide where your chicken coop and chicken yard will be. When making that decision, keep in mind that you will need ready access to a water hydrant and electricity. Running a heavy-duty extension cord along the ground will work temporarily if it has to, but harsh weather, driving, lawn mowing, and other activities will soon make it more of a hassle than anything else.

Once you decide where to house your chickens, you need to decide how to house them. You will need a chicken yard and a coop with nesting boxes that are up off the floor, an area large enough to hold feeders that they can all get around at once, and a place for their waterer. The coop also needs to be one that can be closed up at night to protect your chickens from predators and cold winds. In addition to the nesting boxes, you will need a roost for your chickens to rest on. The roost is usually a wooden pole or 2x4 attached to the walls approximately three feet off the floor.

Before we go any farther with the coop details, let's take a minute to talk about the chicken yard. In your reading and research prior to deciding to raise chickens, I'm sure you came across the term, 'free-range' at least a few hundred times. It is a term that is open to interpretation and can mean anything from letting your chickens roam wherever they want 24/7 to giving them continual access to a fenced in area, opposed to keeping them confined in crate-like cages their entire life.

No matter what term you use, I honestly and wholeheartedly believe it is best to give the chickens constant access (except at night) to a fenced in chicken yard instead of letting them go wherever they choose. Here's why:

- Predators. Your chickens will be safer from predators if you can put them up in a shelter at night. They will also be safer from predators during the day if you have a chicken yard that is (at least) semi-covered.

- Management. You will be better able to manage and keep track of your chickens if you know where they are.

- Efficiency. It takes more time to find the eggs if they aren't in laying boxes or the chicken yard. In fact, often times you won't find them at all—or until it is too late to use them.

We will go into more detail in the next chapter on fencing off the area you will use. For now, let's talk about your chicken coop, also known as a chicken house.

Your focus here should be quality. I'll follow that statement up by saying (bluntly, but unapologetically) that a lot of the chicken coop kits you see in large home and garden stores or even farm supply stores are a waste of money. They aren't durable and don't last. Your best option is to build your own using high-quality lumber and chicken wire.

There are numerous books and websites that can provide you with detailed plans on how to build a durable and functional chicken coop. If you aren't handy and don't feel comfortable taking this project on, you might consider hiring a group of FFA (Future Farmers of America) or 4-H young people to build it for you—with adult supervision, of course. Or perhaps you have a friend or family member who can help you.

Another excellent option is to purchase a ready-made outdoor shed and make some adjustments to it. Just make sure the shed you purchase is made of wood (tin or galvanized metal will make the structure too oven-like during the summer months). The structure also needs two windows, ceiling or ridge vents, and a double door.

NOTE: The double door will make it easier to get feed sacks, bales of straw or pine shavings, and a wheelbarrow inside.

Your shed/chicken coop should be large enough for you to build a wall made of chicken wire stretched tightly over a 2x4 frame the full length of the shed, 'cutting' the shed in half (approximately). This 'wall' also needs to have a single-width in it, so you can go in and out easily to feed and water the chickens, clean the floor and nesting boxes, and gather the eggs.

The chickens' part of the shed should NOT be the part you step into when you enter the shed. The area you step into when you walk through the doorway of the shed should be used to store feed, bedding, vet supplies, and cleaning supplies.

You will also need to cut a small door in the chickens' part of the shed that can be raised and lowered using a pully. This door should be approximately six to eight inches off the floor and have a ramp on the outside for the chickens to go up and down, making it possible for them to come and go as they please.

The chickens will soon get in the habit of going in once it starts getting dark, so you can shut them in to keep them safe for the night.

As previously mentioned, you will need to make sure you have easy access to electricity and water. Chickens need several hours of light a day to prevent molting (loss of feathers) and to regulate and stimulate egg production. Young chickens also need light to keep warm.

Now that you know what you will need for chicken housing, it is time to decide what kind of chickens you want. Some breeds are more aggressive egg producers than others. Some are better for butchering than others, because of their size and meatiness. Some are 'fancy' breeds used for show and gaming activities. Knowing what your purposes are for raising chickens will help you decide what breed or breeds to get.

The next question you will need to answer is how many chickens do you need? Actually, this question needs to be considered when deciding how big to make your chicken yard and coop. Overcrowding is a recipe for disaster! Disease and fighting are the two main problems you will encounter if you have too many chickens in too small a space.

Once you are certain you know…

- where your chicken yard and coop will be located
- how big they need to be
- what materials you will use for fencing and construction
- how much it will cost
- whether you are going to build it yourself, have it built, or buy something pre-built and make the necessary adjustments to it

and have decided…

- what kind of chickens you are going to raise
- how many you are going to start with

it is time for you to get to work.

I know how tempting it can be to bring your chicks home before you have everything in place, but please, oh please, resist this temptation. Jumping the gun almost always causes problems, so it's better to wait and do it right.

Chicken wire

Ready…Get Set…

Chicken coop

How to Raise Chickens

Chicken yard with feeder and waterer

Chapter Three
Ready…Set…and now Go!

Chicken yard securely built? ✓

Chicken coop built? ✓

Supplies purchased? ✓

Good! Now it's time to bring your chickens home.

Where I come from (the Midwest), chicken hatcheries have a selling season during which the vast majority of their baby chicks are hatched out and sold. This season usually runs from early to mid-March through late May or early June. This means the majority of chickens sold to people like you are sold during this time.

NOTE: Knowing when you will be able to get your chickens (if you are starting with newly-hatched chicks) will motivate you to stay focused on getting the chicken yard and coop ready.

Nearly everyone I know starts out with baby chicks. Personally, I think this is the best way to get started, because:

- Starting off with baby chicks gives you and your children the experience and satisfaction of going through the growth process and watching/helping them mature.
- It's easier to transport baby chicks than mature chickens.
- You know how old the chickens are and will reap the benefits of their productivity.

NOTE: This isn't to say buying more mature hens is always bad. For example, I know a young lady who raises chickens as one of her 4-H projects. As part of her project, she does comparisons on how quickly different breeds mature (start laying), the size of the eggs, their hardiness, and so forth. Because of this, she doesn't have the space or the need to keep them very long after they start laying eggs, so she almost always has hens for sale that have been laying eggs for a couple of months.

Buying more mature hens under these conditions would be perfectly fine. But buying them at a swap meet, farmer's market, auction, or from someone who has placed an ad online or in the paper is usually not a good idea. You just don't know what you are getting.

Buying your chicks all at once also means:

- Your chickens will all be the same age and grow up together, keeping the pecking order issues to a minimum.

- You will have a consistent egg supply.

Bringing your chicks home is exciting. They're just so darn cute. What's not to love? But all that cuteness needs specific care and attention. You can't simply take the chicks out of the box you bring them home in and call yourself a farmer. When you bring your chicks home, you need to make sure you:

- Put them in a large round wading pool, stock tank, or even plastic storage tubs—anything that is 12 to 18 inches tall (to keep them from escaping). Line the bottom with a layer of pine shavings, straw, or shredded newspaper.

Ready…Set…and now Go!

NOTE: The container(s) need to be such that they allow about ¼ square foot of space PER CHICK.

- Have a small feeder shallow enough they can reach the feed easily and a small waterer available 24/7.

- Have a thermometer and light. A 'trouble light' works best, as it radiates heat well. The thermometer is ESSENTIAL, as it tells you when to adjust the light's distance from the chicks to make sure they are warm enough, but not too hot.

NOTE: Baby chicks should be kept in an environment where the temperature is 90 degrees the first week or two, 85 degrees for week three, 80 degrees for week four, 75 degrees for week five, and at week six, they should be able to be turned out full time with the temperature being between 65 and 70 degrees consistently (day and night).

NOTE: The temperatures necessary for raising baby chicks is another reason to start in the spring in many areas of the country. By the time they are laying (fully mature), they will be able to tolerate colder temperatures but will still need heat in extremely cold temperatures. The lights you use in your chicken house, combined with their tendency to huddle together, will usually be enough to get the job done.

- Chicks will need the heat and light 24/7 during the first five to six weeks (until you turn them out in the chicken house and yard).

- You will need to monitor them several times a day for health concerns, smothering, and pecking.

Chicks are not so different from any other babies—they require the most care in the earliest stages of their lives, but once you get past the infant and toddler stages, you will be able to enjoy the low-maintenance aspects of raising these beneficial and profitable birds.

Chapter Four
What Chickens Eat

Baby chicks need chick grower feed you can purchase from your local feed store. It is a coarsely ground 'meal' containing the necessary vitamins, minerals, and proteins to promote optimal growth and health. The grower feed also contains essential medications. For those of you who don't want to use additives such as these, check out the websites and other resources at the end of the book for natural feeds and remedies.

Once your chicks mature to the stage of being turned out full-time, you will want to graduate to a pellet or crumble that is formulated to promote egg production (or body mass, if you are butchering your chickens). Chicken feeds are comprised of calcium, protein, salt, and other essential minerals and vitamins.

In addition to chicken feed, chickens LOVE fruits and vegetables. They are especially fond of pumpkin, melon, lettuce, apples, strawberries, tomatoes, and peelings from cucumbers and potatoes (as long as the potato peelings aren't green).

Some people also feed their chickens cracked or coarsely ground shell corn and oyster shells instead of crumbles, along with a steady diet of fruits and vegetables. This is fine, too, as long as they are getting protein from pumpkin seeds and plenty of worms. Most of the time, though, it's easier to feed them the pellets or crumbles to ensure they are getting what they need.

One of the first questions a novice to chickens will ask, is "How much should I feed my chickens? Can I overfeed them?"

Chickens like to eat small portions throughout the day and are naturals at regulating their own intake. In other words, they won't overeat, but they like to eat multiple meals a day, so you need to make sure they always have CLEAN feed in the feeder. As for the table scraps (peelings, rinds, and overripe fruits and veggies), you can give that to them once or twice a day. Don't worry if you skip a day or two now and then. They'll be fine. Just be sure they don't run out of feed.

The weather also plays a role in what you feed your chickens. For example:

- When it is hot outside, chickens enjoy frozen berries and melon. The cold treats keep them from overheating.

- When it is hot outside, you need to limit or cut out cracked or ground corn. The starch in corn produces body heat they don't need.

- Feeding corn and cooked oatmeal in the winter will help generate the body heat chickens need to stay warm.

It is best to keep the chicken feeders containing the pellets or crumbles in the chicken house. This way you won't have to worry about it getting wet. Do keep it out from the nesting boxes and roosts, though, to help cut down on the chances of it getting pooped in.

Table scraps should be fed in the chicken yard. Doing so cuts down on having mice infiltrate the chicken house, as well as flies and gnats, which can cause health problems for your chickens.

NOTE: Chickens are not picky eaters, but there are some things that should not be included in their diet. Don't feed your chickens avocado, citrus fruits, rhubarb, beans, raw eggs, onions, chocolate and other candy, foods containing sugar, or bones.

Equally important to having a constant source of food available is having a constant source of clean water. Hydration is important to all animals, but especially chickens.

You should have a large enough waterer for about half the chickens to get to at one time. They won't all want to drink at once, but you don't want them to have to stand in line. You also want to make sure the water is cool in the summer and tepid or slightly warm in the winter. This helps with controlling body heat.

You will be able to keep the water cooler in the summer if you keep it in the chicken house and/or in the shade outside. In fact, it's not a bad idea to have two waterers for the chickens when the weather allows (no risk of freezing).

In the winter, the water will need to be kept inside the chicken coop. You will most likely need a heater for it. The easiest and most reliable heaters are those you set the waterer on. They are usually self-regulated as far as the temperature is concerned and are equipped with a heavy-duty cord for safety.

As I said at the beginning of the chapter, feeding your chickens a balanced diet isn't difficult. Feeding them the right amount of food isn't difficult, either. In fact, feeding your chickens is easy and economical, which is just one more reason to raise them.

Chapter Five
Chicken Care 101

This chapter is going to follow the good news/bad news/good news format. First we're going to talk about how easy it is to care for your chickens. Then we'll move on to potential problems you can encounter when raising chickens, including health problems, problems within the flock, and management problems. We'll also go over preventative measures and fixes for when prevention isn't possible. Finally, I'll finish the chapter by reminding you that it's better to be prepared and not need the skills, knowledge, and supplies you have than to need them and not have them.

Caring for your chickens is easy as long as you remember these two words:

- Routine. Chickens, like most animals (including us), are creatures of habit.
  - Their routine will generally be the same one day to the next, so if *you* establish a feeding/observation/egg-gathering routine, you will be less likely to miss something.
  - A potential problem can usually be nipped in the bud before it has a chance to do any real damage to your flock if *your* routine makes you well-acquainted with *theirs*.
- Consistency.
  - Make sure you provide a consistent temperature for your chicks as outlined in Chapter Three. Keeping the temperature

of their environment consistent and correct is a matter of life or death.

- Be consistent with feeding times and the amount of feed you place in the feeders. Chickens self-regulate their feed, but if you don't have regular feeding times and aren't consistent with the amount of feed placed in the feeders, your chickens will assert dominance when they are fed.

- Keeping the chickens calm and socially non-aggressive is key to keeping several health problems at bay.

- Be consistent in putting your chickens up at night. Working hand in hand with routine, a consistent 'bedtime' will keep you on your toes against predators.

- Consistently gathering eggs allows you to supply your egg customers with what they want. If you can't keep up with the demand, you will lose your customers. Wasting your product isn't good business.

- If you are letting your hens sit on their eggs to hatch them out, consistently checking them (without being too intrusive) helps you keep track of how many chicks you can expect and should prepare for.

See? That's not so hard, is it?

Chicken care essentials

If you own chickens, you need to have a vet kit containing the following items. Your vet kit needs to be readily available at all times.

NOTE: You need to keep your kit in your house vs. in the barn or chicken coop, so items will be room temperature.

- VetR$_x$ Respiratory and skin remedy (A must!!!)
- Epsom salts
- Disposable gloves
- Aspirin
- Animal wound spray
- Non-stick gauze pads
- Vetrap bandage
- Vitamins and electrolytes
- Corid Solution
- Tweezers
- Scissors
- Dog nail clippers
- Styptic powder
- Flashlight

Predators

Unless you neglect your chickens' housekeeping and dietary needs, keeping your chickens safe from predators will be your biggest concern. The most common predators are:

- Dogs
- Cats
- Coyotes
- Raccoons
- Hawks
- Owls
- Possums (eggs)
- Rats
- Snakes (chicks and eggs)
- Foxes

Prevention by safeguarding your chickens from predators is the best way to deal with this problem. You can prevent predator attacks by doing the following:

- Keep your chickens in a chicken yard and chicken house.
- Use the right materials to keep most predators out.
  - Chicken wire keeps most predators out of your chicken yard.
  - Attaching the chicken wire to boards at the ground and floor level of your coop and yard is nearly essential. Doing so keeps animals from digging their way in.
  - Make sure the gate latches tightly.

- Add a row or two of barbed wire to the top of the fence around your chicken yard to keep some predators from climbing in.

- Don't build the chicken yard around any trees, as owls, hawks, and other predators can use the tree to gain entrance to the chicken yard.

- Make sure you can shut the chicken coop up at night.

• Keep feed sacks in tightly closed bins and get rid of dirty feed so rats and mice won't invade your chicken coop.

• Put your chickens up at night.

• A bamboo windchime deters predators without disturbing the chickens.

• Collect eggs daily to keep snakes, possums, and other egg-sucking critters away.

• Cover your chicken yard if possible. If it isn't possible or feasible, make sure chickens have constant access to the chicken coop. This allows them to escape some predators.

Pests

The most common pests that affect chickens are:

- Flies
- Ticks
- Lice

- Mites
- Gnats

Flies: The best way to deal with flies is to keep your chicken coop clean and hang fly strips in it. Make sure you hang them from the ceiling, so the chickens don't get into them. Keep your chicken coop clean, i.e. clean out the litter (poop) once a week, keep the straw or shaving you use on the floor and in the laying boxes clean, and don't leave table scraps in the chicken yard to spoil and rot. Get rid of what they don't eat before giving them more.

Ticks, mites, and lice: Chickens eat ticks without incurring any harm. But ticks that escape being eaten can pose a threat to your chickens. Mites and lice aren't in their diets, but all three of these insects think a chicken dinner is a tasty treat.

It's impossible to see ticks, mites, and lice that attach themselves to your chickens, but you can make your chickens less appetizing to them by dusting them with a prepared tick, lice, and mite powder. You can also make your own by mixing the following things together: fine sand, dirt (powder-fine), dried lavender buds or crushed dried lemon balm leaves, diatomaceous earth, and ashes from the fireplace. (https://www.backyardchickencoops.com.au/blogs/learning-centre/5-things-add-chicken-dust-bath)

Using dirt and sand may seem counterproductive, but it's not, and the chickens know that. They roll in the dirt and use their feathers to work it down to their skin in order to smother and kill the ticks, mites, and lice that are trying to take up residence on them. Smothering these pesky blood-sucking creatures is an instinctive hygiene and healthcare measure.

If your birds are infested with any of these insects, a dust bath may not be enough to take care of the problem. You will need to treat the entire flock by spraying them with an insecticide for animals. You can also use the 'pouring' method, which means distributing a specific amount of the liquid onto each bird by pouring it along the top of their back. Most of these treatments have not officially been approved for use on chickens, but with a vet's supervision, I assure you they are safe and effective. I'm not purposely trying to promote one over the other, but Ivermectin is extremely effective and I've not known of anyone having a negative experience with it.

Ticks, mites, and lice love to infiltrate the coop, nesting or laying boxes, and the crevices in the floor and wall of the coop. You will know this is the case if your chickens are reluctant to go into the chicken house—especially at night when they will be left there without a means of 'escape.' They know they will be attacked and that intense itching will ensue.

I've said it before, and I'll say it again (and again and again and again)—prevention is your best defense against these pesky parasites as well as all other health and hygiene problems. Cleanliness and neatness provide a strong dose of prevention in dealing with these external parasites. Regularly cleaning out the floor and boxes of the coop is a great deterrent. But that's not all you need to do. At the onset of spring and then again in mid-summer and early fall, remove all bedding, feeders, and waterers from the chicken coop. Spray it down with a solution of bleach water, making sure you lightly saturate the entire surface (including inside the nesting/laying boxes and the roosts), then let it dry thoroughly before putting down clean shavings or straw on the floor and in the boxes. Keeping the chickens out of the coop while you clean and let things dry out is a must, so make sure you choose a warm, sunny day for the job. This will also help things dry quicker. Regularly clean your feeders and waterers with a solution of bleach water, but make sure you RINSE THOROUGHLY.

Coccidiosis

Coccidiosis is a common ailment among chickens. It is caused by internal parasites attaching themselves to the stomach lining. Feeding off the stomach lining causes bleeding in the stomach, which leads to the inability of the chicken to absorb the nutritional value of their feed.

Coccidiosis usually occurs in younger chickens (under six months), but adults are not immune. It is usually only seen when conditions are warm and humid.

The symptoms of coccidiosis include:

- Dark brown or bloody stool
- Ruffled feathers
- Huddling
- Loss of appetite
- Lethargy
- Loss of color in their comb

NOTE: The symptoms of coccidiosis are similar to many other chicken diseases, so it is difficult to diagnose. But since it is easy to treat and most chickens recover from it (unless you wait too long to address the symptoms), treating for coccidiosis is a good first step to take when these symptoms present themselves.

The treatment for coccidiosis is relatively simple: add Corid (amprolium) to their water. Again, this solution is marketed for cattle, but is completely safe and effective for use in treating other species of livestock. Farmers across the world use it to treat chickens, sheep, cattle, goats, and hogs with no problems.

You will need to add ½ teaspoon of liquid Corid per gallon of water every day for five days. If you use the powdered Corid, the dosage is 1/3 teaspoon per gallon of water for five days. Make sure you mix a fresh batch every day for those five days and keep things extra clean.

Coccidiosis can spread quickly, because the parasites are present in the feces, as well as the stomach. Chickens who walk in an infected bird's feces and then preen themselves, or that peck in food contaminated by the infested feces, then become infested themselves. Because the disease is communicable, you should remove sick birds from the flock, but treat all birds.

You should also follow up this treatment by making sure you add some vitamin A and K to their diet.

Conjunctivitis

Conjunctivitis is an eye ailment or irritation. It is caused by excessive ammonia fumes emitted from urine and feces. It is not contagious, though it might seem that way since it usually affects more than one chicken in the flock.

The symptoms of conjunctivitis are watery eyes, cloudy eyes, chickens rubbing their eyes with their feathers, and not wanting to be in the sunlight. If not dealt with properly, blindness can occur.

Preventing and treating conjunctivitis is simple. Keep the chicken coop clean and make sure you have proper ventilation. If you don't, and your chickens contract conjunctivitis, they will recover in a month or two after you do your part. They will recover, that is, as long as it didn't cause blindness. There is no coming back from that.

Campylobacteriosis

This is a fairly common disease affecting only chickens. It is bacterial and is spread through contact with the poop of infected chickens. Flies, roaches, and mice can carry the bacteria into the flock after coming into contact with infected chickens—even in another flock. People can bring it into a flock on their boots (walking in infected chicken poop from one coop and bringing it to another).

The symptoms of campylobacteriosis vary, depending on the age of the chicken. Chicks affected by these bacteria are slow-growing and have diarrhea. They will also be lethargic. Mature chickens usually die suddenly, despite appearing healthy. But upon further examination, you might realize their egg production is down, and/or their comb is scaly and small.

There is no cure for this bacterial disease and the only way to prevent it is to have excellent coop and chicken yard management and sanitation practices.

Breast blister

Breast blister is exactly what it sounds like—a blister on the breast of the chicken. It is caused from rubbing against rough surfaces such as the roost, wire fencing, or nesting boxes. It can also come from the pressure of being up against these surfaces if the chickens are too large for the space.

It is not contagious, and prevention is easy. Make sure your surfaces aren't too rough and make sure your nesting boxes and yard area can accommodate the size of your chickens comfortably.

To treat the blister, drain it, treat it with iodine, and apply a thick layer of triple-antibiotic ointment. You will need to clean and re-apply the ointment daily until it begins to heal over.

Cholera

The word 'cholera' sounds scary, doesn't it? And with good reason. This disease, which is unfortunately common in chickens, can cause great harm to your flock. It affects flocks that are free-range and live in warm, moist climates more so than other flocks.

Cholera is a bacterium transmitted from one bird to another (mucus or eye discharge) and is contracted through contaminated feed or water. The symptoms of cholera are lack of appetite, diarrhea, lethargy, ruffled feathers, excessive thirst, fever, increased respiratory rate, paleness, mucus from the nose and mouth, and 'sudden' death.

I put '' around the word sudden, because your chickens will be showing other signs of infection prior to their death, but because so many of the symptoms chickens display are symptomatic of several different illnesses, it can sometimes be difficult to make an accurate diagnosis right off the bat.

There is no cure for cholera and any birds infected or showing possible signs of infection should be removed from the flock immediately. If you know for sure the birds are infected, you need to destroy them and immediately start disinfecting any and all potentially affected areas and surfaces. This is nothing to play around with.

Anemia

Anemia in chickens means basically the same thing as it means in humans. It is blood lacking in the proper nutrients. The symptoms of anemia in chickens include pale skin, combs, and wattles, lack of energy, ruffled feathers, huddling together (possibly because the poor circulation results in a need for shared body heat), and sudden death in birds that seem otherwise healthy.

Sometimes anemia in chickens (and other species of livestock) is caused by internal parasites called barber pole worms. These worms feed off your chickens' blood supply until they are depleted to the point of not having enough blood to pump through their little bodies.

The best way to combat anemia not caused by parasites is to make sure your coop is clean and well ventilated and that they are receiving a sufficient amount of sunshine (vitamin D) and vitamin K in their feed.

Preventing anemia caused by parasites is best done by worming your chickens using a ready-made chicken worming product such as Safeguard or Panacur. Better yet, VetRx can be used to prevent this type of infestation (as well as preventing and treating many other problems).

NOTE: Chickens kept in close quarters are much more apt to develop this problem than those that are allowed to run in a chicken yard and come and go from the coop of their own volition.

Bronchitis

Bronchitis is an infectious disease that causes similar symptoms in chickens as it does in humans—sneezing, coughing, raspy breathing, and runny nose and eyes. Most of the time, bronchitis is treatable and doesn't last long. However, that depends on how vigilant you are and whether or not your management practices are such that you catch the problem early on.

The best treatment for bronchitis is good ventilation, electrolytes in their water, a little extra food, keeping the temperature in the coop comfortably warm, but not humid, and making sure you aren't overcrowding your chickens.

Malaria

Malaria is carried by mosquitoes and affects the blood. There is no cure or treatment for malaria in chickens. If they are infected, they will die. Prevention is key for this one. Martin houses (martins are a type of bird that are especially fond of mosquitos) keep the mosquito population at bay, as does having a few guinea hens. You also need to make sure to not invite mosquitos to your farm via stagnant ponds, rain barrels that aren't covered, and other things or areas on your property where water collects and stands. Keeping the air circulating in your chicken coop using an oscillating fan also helps keep the pesky little blood-suckers away. You might also want to take the time to spray with Malathion, which is proven to be highly effective.

Pox (wet or dry)

Pox in chickens (not exactly like chicken pox) is another fairly common ailment that can affect your flock. Dry pox causes wart-like bumps on the skin, legs, waddle, and comb. Wet pox is a respiratory version causing scabby bumps to grow on the face and eyes, and in the windpipe and throat.

Pox is caused by a virus spread through contact between chickens, and through mite and mosquito bites. The best preventative measures you can take against either type of pox is to control the mite and mosquito populations in your coop and chicken yard, and to practice good management and sanitation.

There is no treatment or cure for dry pox. You might have to scrape the bumps off from around their eyes and mouth so they can eat and see better. If you do, apply a bit of antibiotic ointment as an added safety precaution. For wet pox, you can do the same. You might also need to swab their throats with a cotton swab dipped in a light iodine or a light VetR$_x$ solution to sooth the discomfort.

Bird flu

Bird flu is something we've all heard of. It is also something everyone who has chickens needs to be aware of and knowledgeable of. That being said, large outbreaks are not all that common, probably because at the first sign of this killer disease, chicken owners take quick and decisive action. It is also important to know that there are several different strains of bird flu, some more lethal than others.

Bird flu normally attacks the respiratory system, but in some cases, the nervous and digestive systems are also affected. The most common symptoms of bird flu include coughing, sneezing, watery eyes, ruffled feathers, huddling, and *green* diarrhea. Some strains present very few symptoms. The affected chickens will just die for no apparent reason.

Bird flu is highly contagious. It is transmitted through feces and other bodily fluids from infected birds. Rodents can also pass the flu from flock to flock.

Don't let anyone that might be a possible means of exposure in your chicken coop or yard. Likewise, don't visit flocks that might be infected, as you could carry it back to your own flock on your boots.

Infected birds should be quarantined and treated with antibiotics. Your vet can tell you which antibiotics will work on the particular strain of flu your chickens have. If your chickens have one of the most dangerous strains of bird flu, you must report it to the federal health officials, as it can affect humans.

There are several other diseases and ailments that might affect your flock to some extent or another. Having a fastidious sanitation routine for keeping your coop, feeders, and waterers clean and free of feces, insects, urine, and other contaminants greatly reduces your chances of ever needing to worry about any of them.

Did you get that? **Excellent management practices regarding cleanliness and observation are the key to a healthy flock.**

NOTE: For a more comprehensive list of illnesses and ailments, along with their causes, symptoms, treatment, and preventative measures, go to: http://www.raising-chickens.org/poultry-ailments.html

Chapter Six
Which Breed is Best for You?

Did you know there are hundreds of different breeds of chickens? Having so many choices might sound like a good thing, but it can also be a bit confusing. They all have some distinctive qualities and different breeds even have different personality (or chickenality) traits. That's why you need to decide what purpose your chickens are going to serve before you buy. Once you decide why you want to raise chickens, you can choose which breed or breeds are best for you.

Why raise chickens?

Chickens have two purposes: meat and eggs. From those two purposes stems a number of reasons people choose to raise chickens. Some of them are as follows:

- Raising some of your own food (self-sufficiency)
- Additional income from selling eggs and/or butcher chickens at farmer's markets, online, and through word of mouth
- Raising for resale, i.e. fertilized eggs for hatching, chicks or young layers, eggs to local bakeries and cafes, fresh chickens (butchered) for local eateries
- Show chickens
- Poultry for hunting (ducks, quail, etc.)

Depending on the amount of space and time you have, you might not have to limit yourself to just one reason. A word of caution, though: don't take on more than you can handle *well*. Problems can arise due to mismanagement and poor sanitation if you have more than you can properly take care of. Remember—it's better to do a few things really well than a lot of things in a mediocre fashion.

Now let's get down to the business of which breeds to consider. There are far too many to list them all, so I am only going to list those which are tried and true, those which are an excellent ambassador of their species.

Chickens that lay brown eggs

- Rhode Island Red-heavy egg layers, hardy in a variety of climates/temperatures, generally healthy and robust, meaty enough to butcher

- Plymouth Rock- heavy egg layers, calm temperament, great for beginners and children, hardy and robust, meaty enough to butcher (often called Barred Rock)

- Buff Oropington- heavy egg layers, calm and friendly, tolerate a variety of temperatures, can be butchered

- Dominique-medium to heavy egg layers, calm and friendly, robust and generally healthy

- Cherry Eggers- medium to heavy egg layers, small but fast-growing, extremely calm, hardy and easy to manage

Which Breed is Best for You?

- Production Red- heavy layers, hardy, easy to raise, a bit on the smaller side, not good for butchering, fairly mild-mannered

- Jersey Giants (white or black)- largest breed, not fast-growers, great for butchering (but rarely used in large operations because it takes too long to get them to size), hardy, solid egg producer, (but again, slower to mature, so it takes longer to start getting eggs)

Other brown egg producers include Delewares, Red and Black Sex Links, and the Cinnamon Queen.

Chicken that lay white eggs

- White leghorns-nothing beats a white leghorn for egg production (white, brown, or colored). These chickens are heavy layers, they mature quickly, and they are big, hardy, and robust. They are confident, so they tend to be a bit flighty. The leghorn rooster is known for being one of the most aggressive roosters.

- Brown leghorns- the brown leghorn is equally prolific in its egg production capabilities. It is similar to the white variety in all other aspects of growth and temperament.

- Polish- there are several varieties of polish chicken (black, white, blue, buff, golden, etc). They are a fancy looking chicken, laying much smaller eggs than other breeds, and a decent amount of them.

Chicken that lay colored eggs

- Easter egg- the eggs are medium-sized and come in a variety of colors (blues, yellows, greens, and pink). They are not nearly as prolific as a leghorn, dominique, or Rhode Island red, but they are consistent and reliable for several years. The Easter egg chickens do have some genetic issues with beak problems, but it won't cause disease or sickness in your flock. It's more of an overbite issue.

- Ameraucana- there are several varieties of this breed of chicken, but they all lay colored eggs. They are among the most expensive breeds of chicks. They also tend to be less hardy than other breeds, meaning they do best in temperate climates.

Meat chickens

Many of the breeds listed in the brown egg section are considered dual-purpose chickens, meaning they work well for both meat and egg production. Choosing a breed that is dual-purpose is what most producers prefer to do. If, however, you want to be a bit more specialized, the Cornish chickens are the way to go.

Other things to consider

In addition to the color, size, and quantity of the eggs and whether a breed is meaty enough for butchering, you also need to take into consideration the following:

- Do the hens make good mothers? If you want to hatch out chicks, will your hens be good setters? Do they have good maternal instincts?

- How aggressive are they? Some breeds are more aggressive than others. If you have children, you need to take this into consideration. Teaching children responsibility is among the most important things you will ever do as a parent. One of the best ways to do this is with livestock, so make sure they have a safe and positive experience.

- If you are going to have more than one breed, you need to make sure they are similar in size and personality, so as not to aggravate the 'pecking order.' And yes, it's a real thing we'll cover in the next chapter.

- If you have enough available space, you can have different breeds, but keep them separated. If you don't have the room to separate them, you can try adding extra feeders and waterers, so no one feels the need to dominate the area.

For more detailed information on breed characteristics, talk to your county agriculture specialist, extension agent, hatchery, or veterinarian. You can also learn quite a bit by visiting the following websites:

- https://www.cacklehatchery.com/

- https://www.mcmurrayhatchery.com/index.html

- http://www.idealpoultry.com/

How to Raise Chickens

Barred Rock

Which Breed is Best for You?

Cherry Egger

Leghorn

Which Breed is Best for You?

Easter Egg Chicken

Chapter Seven
Pecking Order...It's a Real Thing

Chickens are social animals. They enjoy the company of their peers. But just like any other animal (including humans), they establish a social ladder or hierarchy. It's their way of keeping order. And to keep order, they literally peck at one another...ergo, the term "pecking order."

The pecking order is a natural process in every flock. It establishes who is the boss and who isn't. It establishes rank and file—a chain of command, you might even say. It influences their eating, drinking, mating, and roosting habits. Hens higher up on the pecking order choose which nesting boxes to roost and lay in. Those lower on the totem pole get what's left over. It even affects when and where they take their dust baths. And that's just the hens. Don't even get me started on the pecking order among roosters!

The dominant or alpha rooster usually proclaims his status by crowing. A rooster will also crow to let the hens know he's in charge of the flock. But that's not always a bad thing. If you have a rooster in your flock (which you don't have to), he tends to be protective of the hens, watching out for and guarding against predators. But on the downside of that, a rooster often sees you as a predator when you come to gather his hens' eggs. ☹

Pecking order in the flock is established early. A group of baby chicks just days old will establish pecking order. This usually takes place around the feeders or the light/heat lamp. The most dominant chicks will bully the weaker ones in order to eat first and stay warm.

They do this by pecking on their heads and backs. You need to be vigilant and mindful, watching carefully for signs that this is happening. As long as it is slight and doesn't cause bleeding, don't worry too much about it. It is nature being nature. Just like 'boys will be boys,' chicks will be chicks.

As the chicks get older, they tend to maintain their dominant place in the flock by bumping the weaker or 'lower' chicks out of the way instead of pecking on them. Chickens quickly learn who they can and cannot push around. Because the pecking order is natural and instinctive, and because it happens quickly, you usually won't have to intervene. As I said earlier, unless there is bleeding involved, it is usually best not to. But if it is obvious there is injury to the chicks or chickens being bullied, you need to remove the injured birds, treat them as needed, and keep them separate from the rest until they are fully recovered. At that point, you can try to reintroduce them to the flock, but make sure you keep an eye on everyone to make sure the same thing doesn't happen again.

As long as the chickens are similar in size and the ones you are reintroducing into the flock are fully recovered, 'round two' of bullying usually isn't an issue. One bird will always find another that it can push around, allowing that one to climb higher in the pecking order.

New chickens and the pecking order

How you introduce new chickens into an existing flock depends on where the newcomers come from. Chicks that are hatched out from sitting hens are usually readily accepted. If a chick or chicks are sick or injured, the others will probably bully them, making it necessary for you to remove them until they are bigger and healthier. When introducing chicks not naturally born into the flock, taking things a step at a time is the key to success. Keeping them separated but visible to the existing chickens will allow your current chickens to get acquainted with the new ones. This way they can start the pecking order while getting to know each other through interaction without actually coming into contact with one another and it will also give you an idea of how they may act once they are together. Keeping the birds separated for a couple of weeks before putting them together is usually all it takes.

NOTE: To make the pecking order process even less stressful to new chickens, provide plenty of hiding places for them, multiple feed and water areas, and plenty of roosts in the coop.

NOTE: If you have a rooster in your flock, he will undoubtedly be at the top of the pecking order. Roosters are very proud animals and the thing they are proudest of is their status in the flock. Often, however, the rooster will view you as a threat to his dominance. Doing the chicken/wing dance around you and charging towards you are signs that they feel threatened by you. Pecking and jumping at you to claw or spur you and attacking you every time you turn your back usually follows, with the bird's behavior getting progressively worse and more aggressive. If you detect this early, you can stop it by making the chicken view you as the dominant one. Here's how:

- Never run from the bird; stand your ground. Shoo or chase him (or her) off.

- If he still tries to attack you, grab him and hold him down to the ground with a hand around his wings and back, and the other hand on his head/neck. Be firm, but gentle. The objective isn't to hurt him, but to restrain him until he is calm. Hold him like this for a minute or two, then calmly release him. You shouldn't have to do this more than once or twice before he realizes he won't be able to dominate you.

- If the behavior persists, or if you have children who might be harmed, the best advice I can give you is to have the overly aggressive rooster for dinner.

Nature at work

As harsh as it may sound, the pecking order is a natural phenomenon. Establishing a hierarchy is something nearly every species does. That doesn't mean you should sit back and do nothing to prevent it from harming or killing your chicks, but it does mean you need to realize that there will be times when your best efforts won't be enough to save a chick (or two or three). The key to success in keeping the pecking order bullying to a minimum is observation. The more observant and proactive you are, the safer your chickens will be.

Chapter Eight
The Purpose of Your Flock

Just like the title of the book implies, chickens are a multi-purpose animal. The most obvious purpose for raising chickens is the fact that they provide two different sources of food for those who want to be more self-sufficient. In addition to food, chickens can also be a source of supplemental income. Here are a few common purposes for raising chickens:

- Selling eggs
- Selling butcher chickens
- Selling laying hens
- Alternative uses for eggs that generate income
- Selling newly hatched chicks
- Selling fertilized eggs to be hatched out in incubators (schools, 4-H kids, homeschool families, hobby farmers)
- Growing your flock
- Producing food for your family

Selling eggs

Selling the eggs your chickens lay is the typical go-to for making a little extra money with your chickens. It is also the easiest way to make money with your chickens because a) if you have a dozen or more hens you are going to have a steady supply of excess eggs and b) you'll have no problem selling farm-fresh eggs.

The number of hens you have will decide who your egg customers are. If you have one or two dozen hens, your egg customers will most likely be extended family members, friends, neighbors, people you work with, your child's teacher, and/or people you go to church with.

Generally speaking, a flock of one or two dozen hens means you will gather two dozen eggs per day. Assuming you keep a dozen or two per week for your own use, that will leave you with five or six dozen eggs a week to sell.

If you have a larger flock (three dozen or more) you will naturally have more eggs to sell. If this is you, you will want to consider one or more of the following options for selling your eggs, because each of these allow you to sell larger quantities in a short amount of time:

- Weekly farmer's market—they are usually five hours or less and have a large customer base.

- Local café or bakery—selling your eggs wholesale to a business allows you to sell all of them at once, meaning you literally put all your eggs in one basket. ☺

- Sell your eggs wholesale to a natural foods grocery store, farm-to-table store, or health food store.

- Contact your state's agri-business division to learn how you can be included in promotions and events they sponsor.

The Purpose of Your Flock

Advertising your eggs can be done via word of mouth, online ads, social media, farmer's market booth, signage in your yard, and going door to door to markets, bakeries, and cafes to discuss the possibility of being a supplier.

NOTE: You will likely need a business license to sell your eggs wholesale. These licenses are very inexpensive and allow you several business-related deductions on your taxes.

NOTE: Selling at a farmer's market or to friends and neighbors out of your home usually doesn't require a license, but check with your local and state officials to make sure.

Butcher chickens and/or laying hens

Selling butcher chickens is, quite frankly, the least popular and least cost-effective means of making income from your chickens. The investment (feeding them) vs. profit just isn't there. And don't even think about selling fresh chicken (those you've butchered). The health regulations are justifiably restrictive, making it extremely inefficient from a business standpoint. Instead, selling live young hens that aren't quite ready to lay to people who want to raise their own meat or egg producers is the only reasonable way to go. You can sell your chickens at farmer's markets or from home.

You can advertise your chickens the same way you advertise for egg sales:

- Word of mouth
- Social media
- Signage in your yard
- Farmer's markets

You also have the option of advertising at:

- Online ag-product websites
- Feed stores
- Animal swaps

Alternative sources for income using eggs

Now we're going to spend a few minutes thinking outside the box, which is one of my favorite things to do. I enjoy coming up with creative ways to use the things I have available to me. And when we're talking about farm products, the term most commonly used is 'value-added agriculture.'

A friend and fellow author is somewhat of an expert on value-added agriculture. She worked side by side with members of her state's agricultural division to create their value-added agriculture program, as well as their agri-tourism program. So naturally, I went to her for some ideas on what a person might do to generate a bit of income from their chickens above and beyond the obvious. Here are some of the ideas she gave me:

- Angel food cakes: making them from scratch is a rarity these days, so people will pay a decent price for one—especially at farmer's markets or when sold in a farm-to-table venue. NOTE: You will need to prepare the cakes in an approved kitchen if selling anywhere besides a farmer's market or most online venues.

- Blow the eggs, use the inside, clean and dry them, and decorate the shells as Christmas ornaments. If you aren't artistic enough to do that, sell them to craft shops or local artists who are.

- Sell quiches at the local farmer's market or, if you can cook in an approved kitchen, sell to a local bakery or coffee shop.

- Crack the eggs and use or freeze the contents for later use. Rinse the shells, fill them with soil, and start seedlings in them to sell. The shell can be planted right in the ground.

- Keep all your eggshells, rinse and crush them, and add them to dried, used coffee grounds, package in small bags, and sell as houseplant fertilizer.

- Make and sell all-natural sidewalk chalk. Pulverize the eggshells, mix with one teaspoon of hot water, one tablespoon of flour, and a few drops of food coloring. Pack the mixture into small silicone molds or an empty toilet paper tube and let dry thoroughly.

Fertilized eggs or newly hatched chicks

Selling fertilized eggs or newly hatched chicks is what you would call a niche market, meaning your customer base is somewhat limited in both volume and frequency. In other words, you won't be able to count on this as a steady source of income. This doesn't mean you should write it off altogether, though. Being able to give others the experience of watching life happen by hatching out eggs is something you can take pride in—especially these days when so many children (as well as their parents) don't know where their food comes from.

If this is something you think you would enjoy doing, the first thing you need to do is make sure you know for certain that you have fertilized eggs to sell and that they are being incubated either by the hen or in an incubator, so that the success rate for hatching will be high.

NOTE: Your reputation and the expectations of children are involved here, so you want to be integrous.

In order to be confident in what you are doing, you need to do the following:

- Have a rooster in your flock that will fertilize the eggs. I know this seems like a given, but you would be surprised at the number of times I've been asked if a rooster is necessary for hatching chicks out.

NOTE: For those of you who don't know, hens store sperm the rooster deposits inside them when mating. Some of the sperm collects in little pouches in the uterus, while some travels toward the oviducts where the egg is starting its decent to be laid. If there is sperm present when the egg passes through the oviduct or uterus, it will most likely become fertilized.

Roosters can mate with up to 15 hens between 15 and 25 times a day. This means that if the hen is cooperative and receptive to her suitor, she can hatch out several chicks of her own over the course of just a few days.

- Prepare your incubator. You can purchase an incubator from farm supply stores (brick and mortar or online). You can also make your own by using a Styrofoam cooler, wooden box, or plastic storage tote, a 25-watt light bulb, duct tape, a thermometer and humidity level gauge, and a bowl of water. Here is a link that shows you how to make your own incubator: https://www.backyardchickens.com/articles/category/incubators-brooders.27/

- Gather your eggs (more on that in a moment) and place them in the incubator and then follow the instructions given below for hatching out the eggs.

You need to go through this process a couple of times and become comfortable with candling your eggs (again, more on that in a moment) before you instruct other people on how to do it. Even if you don't want to bother with selling fertilized eggs or newly hatched chicks, yet want to hatch out your own chicks, you still might want to try incubating them instead of, or in addition to having the hens hatch them out. Either way, here are the basics of hatching out eggs. For more detailed instructions, you can visit the website given at the end of this section.

1. Select undamaged eggs for incubating. The fresher, the better, but if the eggs are four or five days old, as long as they have been in the nest and the temperatures aren't too cold, they will still probably hatch.

2. Don't wash the eggs. Washing the waxy coating (bloom) off the egg exposes its pores to bacteria. Also, place the eggs in the

incubator so that they are laying in the same position they would be if you laid them on the kitchen table—with the larger end of the egg a little more elevated than the smaller end. The reason for this is for proper development of the embryo and then later on, so that the developing chick doesn't drown in its own fluid.

3. When placing the eggs in the incubator, make sure the incubator's temperature is maintained at 95 to 100 degrees F. You will also need a hygrometer to measure the humidity level in the incubator. The desired humidity level is 25-50% for the first 18 days, then 65 to 75% until the chicks are hatched. You will need a bowl of water in the incubator to help maintain the proper humidity level. The humidity level is important for proper development of the egg's shell. If the humidity is too high, the shell will be too soft and the chick will ingest fluids into its lungs and drown. If the humidity level is too low, the shell won't crack and the chick will die inside.

4. Once the eggs are in the incubator, they will need to be turned at least three times a day. Most people experienced in hatching eggs out say the number of times you turn your eggs each day should be an uneven number so that the eggs never spend the night on the same side two nights in a row. The reason for turning the eggs is so that the embryo doesn't get stuck to one side of the shell for an extended period of time. If this happens, the chick will grow abnormally, causing all sorts of problems.

5. Once the eggs have been in incubation for four to seven days, you will need to 'candle' them to make sure they are fertilized.

Candling means using a bright light to look through the egg's shell to see what's inside. This must be done in a room that is completely dark. You simply shine the light on the egg. Make sure you are holding it in such a way that the light can shine all the way through. The best way to do this is to use your thumb and first finger to hold the egg from tip to tip.

6. If after day ten you see nothing, the egg needs to be discarded.
7. It takes between 21 and 25 days for eggs to hatch. If you have cared for your eggs properly during this time, you can expect a 75% hatch rate.

Once the chicks are hatched, give them a few hours to dry themselves off and acclimate to the world. Some people leave the chicks for a couple of days. This is fine, as they can live off the nutrients they received while still in the egg for a couple of days. Do be careful, though, that they don't drown in the water you are using to maintain proper humidity levels in the incubator. When you take the chicks out, care for them as you would the chicks you bring home from the hatchery.

Food for your family

Last, but certainly not least, you may choose to raise chickens for the sole purpose of providing food for your family. If this is the case, you will need no more than a dozen hens and no rooster will be necessary. Even at that, you will undoubtedly have more eggs than you need throughout most of the year. That's okay, just use some of the alternative income ideas you just read about to 'pay it forward' with random acts of kindness for friends and neighbors.

No matter what you decide, one thing I am certain of is this: you won't be disappointed you added chicken raising to your resume.

The Purpose of Your Flock

Candling Day 4

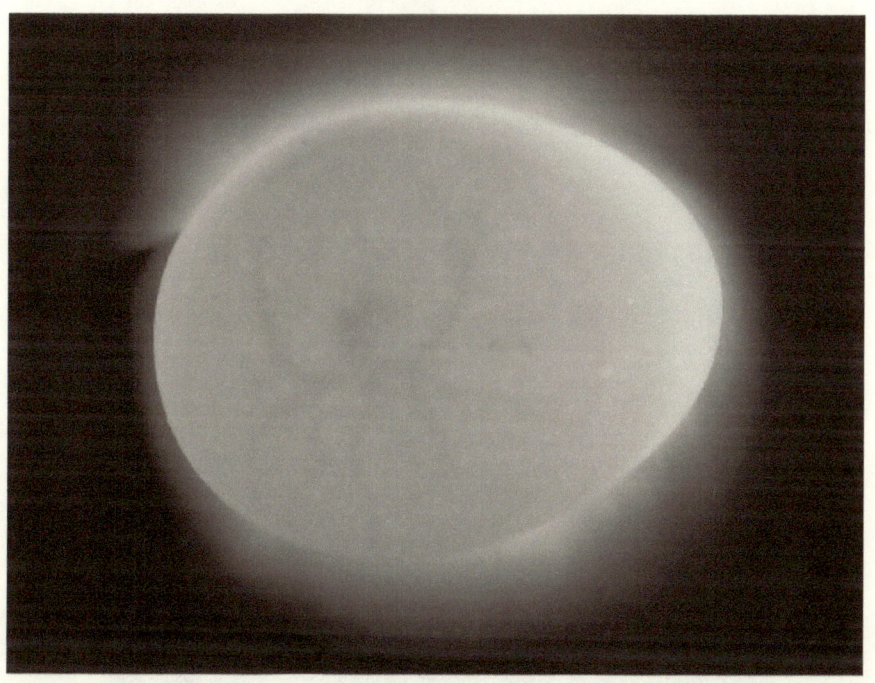

Candling Day 10

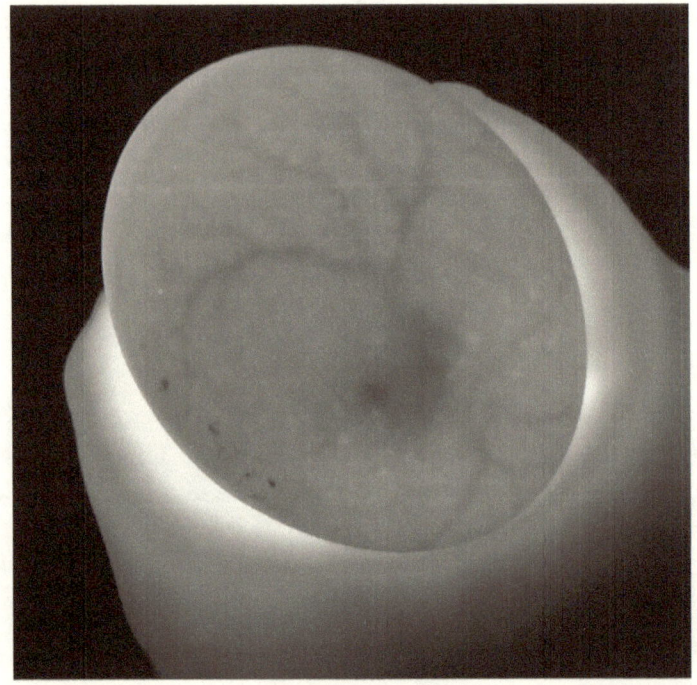

Chapter Nine
Questions and Answers

At this point we've covered the important basics of how to care for your chicks, what to watch for in the way of health concerns, how to care for your mature hens, what chickens need in the way of 'room and board,' and how to make some extra income from your chickens. This is all important information you need in order to be successful, but experience tells me you still have a few questions. So, to finish out our time together, I want to answer the questions most often asked.

NOTE: Some of this information may be a repeat of what you've already read, but it never hurts to be reminded of these things.

Q: How much room do I need to raise chickens?

A: The rule of thumb is that a chicken needs 15 square feet to strut around outside. That means if you have a dozen hens, you need approximately 200 square feet of space for a chicken yard. To help you visualize that, the average living room is 16x16 (256 square feet).

Q: How big does my chicken coop need to be?

A: People's opinions on this vary greatly. I believe as long as every hen has a nesting box, there is a roosting pole large enough to accommodate all the hens at once without crowding, the area is large enough that all the hens can gather at the feeder(s) without feeling crowded, and the feeder(s) can be situated so that hens can roost without poop dropping into the feeders or waterers, it's big enough. Some people choose to have long-run coops, meaning the feeders and waterers are on one end, with the roosting bar and nesting boxes on the other. Long-run coops are usually just like the name implies—somewhat narrow (6-8 feet wide), but long (20 feet or more).

Q: How expensive is it to feed chickens?

A: Nearly mature and mature chickens should be fed the equivalent of about ½ cup of feed per day. Feed prices vary according to brand and several other factors, but on average, a 50-pound sack of egg-layer feed with probiotics to promote optimal health runs about $15. That means a dozen hens would consume that 50-pound sack of feed in approximately eight or nine days. This is not taking into consideration the other things you can feed them (table scraps), so generally speaking, you can plan on buying a sack of feed every two weeks. That comes out to approximately $30 a month.

NOTE: Chicks eat a lot for their size, but 'a lot' is a relative term. And since you will undoubtedly throw out a portion of the food you put in your chick feeders because they poop in it, it's hard to predict exactly how much you will spend on their diet. Once they get older, they tend to stop pooping in their feed (as long as you keep the feeders away from nesting boxes and roosting poles).

Q: How much money can I expect to make selling eggs?

Questions and Answers

A: There are a lot of variables involved this question, so to answer it, I'm going to give you an example you can work from to tailor it to your specific situation.

For the sake of time and space, we'll say that the time and money spent raising your chickens from chicks to laying hens is your business investment. If you really want to get it down to the penny, you can keep track of all your expenses during this time period and figure them into the equation when you are doing this for yourself.

If you have a dozen hens, you can expect them to lay an average of 75 to 80 eggs a week (that's giving the girls a little grace ☺). Let's say you keep two dozen eggs per week to eat. That leaves you with about four dozen eggs to sell per week, or 16 a month.

Depending on where you live, most people will pay $2-5 a dozen for farm-fresh eggs. That comes out to about $56 a month, which is a profit of $26 a month. No, it's not going to make you rich, but if you use fewer eggs per week yourself, and if you explore the value-added opportunities, you can increase your profits.

Q: How long do chickens lay eggs?

A: This chart gives a fairly accurate explanation. Of course, this will depend on the breed, the care and feeding regimen, and the weather and amount of light they are exposed to.

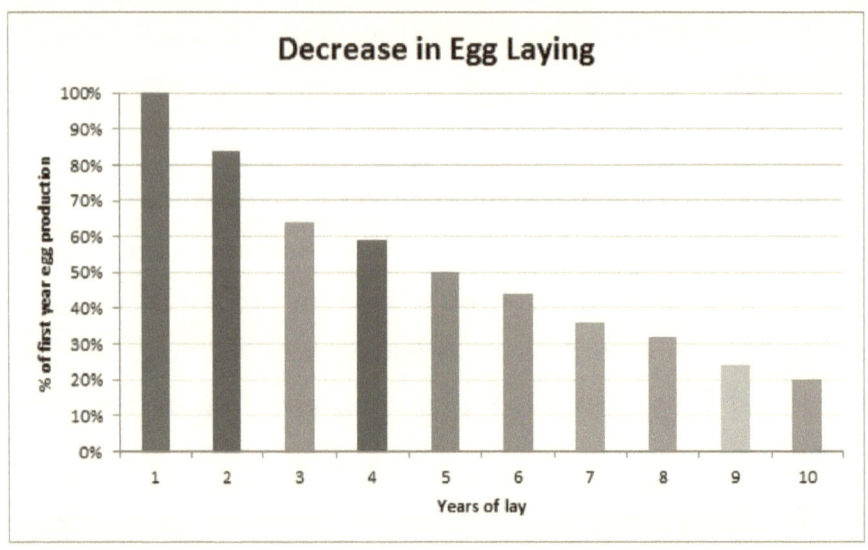

Q: What do you mean when you say 'approved kitchen?'

A: An approved kitchen is one your county health department deems suitable for cooking and preparing products to be sold commercially. Often small producers (like yourself) will pay a nominal fee to use a church or preschool kitchen after-hours, or become part of a co-op with other producers.

Q: Can I file the Schedule F farm tax form?

A: Yes, if you meet the qualifications. They are stringent and you don't have to be a large producer or even make a profit to do so. As long as you show income from your farm (even if the expenses are greater than the income), you can file. You do, however, need to have proof of all expenses and income. The IRS provides all the information you need to file the Schedule F. Your local farm service agent will also be able to answer any questions you might have.

Questions and Answers

Q: There are no hatcheries close by. Where do I get chicks from?

A: You can ask your local feed store for names of people who have chickens. Contact them to see if they hatch chicks out or try your hand at hatching them yourself. You can also order chicks from hatcheries and they will mail them to you.

Q: Where do I get feed and supplies for my chickens?

A: Your local feed stores, farm stores, and mail order or online suppliers can provide you with everything you need. To find your local feed stores, contact your local farm service agent, county extension office, or search for them online or in the (gasp) phone book. For things they may not carry, you can visit any of the following links to find what you need:

https://www.jefferspet.com/pages/poultry

https://www.premier1supplies.com/poultry/species.php?msclkid=020ed5f5c3031d77a1f0892c27956177&utm_source=bing&utm_medium=cpc&utm_campaign=(ROI)%20Equipment%20%26%20Supplies&utm_term=chicken%20supplies&utm_content=Chicken%20Supplies

http://www.poultrysupplies.org/

https://www.farmtek.com/farm/supplies/cat1a;ft_poultry_equipment.html

https://www.valleyvet.com/c/livestock-supplies/poultry-health/poultry-equipment.html?ccd=IMS007&grp=1000&grpc=UUUU&grpsc=UUUU&catargetid=120295250000402522&CAPCID={CREA

TIVE}&cadevice=c&catci=dat-2333301113114029:loc-190&agid=1304020894886584&msclkid=10ef69624e6411bba29935bbaac3bf65&utm_source=bing&utm_medium=cpc&utm_campaign=%5BADL%5D%20%5BNon-Brand%5D%20%5BEC%5D%20%5BUS%5D%20Dynamic%20Division%20-%20Farm%20-%20Livestock-Supplies%20Category&utm_term=livestock-supplies&utm_content=Dynamic%20Livestock-Supplies%20Categories

Q: How can I get my kids involved in caring for the chickens?

A: Make it their responsibility to feed, water, and gather the eggs. Make sure, however, that they know how to do it properly. You don't want to set them up to fail or cause unnecessary stress and harm to the chickens. You can also require older children to help clean the coop. A little (or a lot) of hard work never hurt anyone.

Children over the age of eight can also be involved in 4-H and choose poultry as one of their projects. Not only will this encourage and require your kids to be involved in the care of your chickens, they will learn a great deal about the animals themselves as well as countless other valuable life skills.

If your children are involved in the care and management of the chickens, you also need to make sure they receive some of the benefits. You can do this by paying them a percentage of the profits you make from your chickens. Or if you don't sell the eggs, pay them an allowance for the work they do.

Closing Comments

In all likelihood, your grandparents, and possibly your parents, know what it means to be at least somewhat self-sufficient. They know what it's like to taste a tomato, warm and straight off the vine. They know that potatoes are dug out of the ground, that cheeseburgers come from cows, and how to get corn out of the husks and onto the table. I understand that the world we're living in today is much more crowded and compact. I get that very few of us have the luxury of living in a place where we can have a few chickens, a big garden, and possibly even a steer to put beef in the freezer. But if you do have the ability to have even one or two of these things, I can tell you that the feeling of accomplishment and satisfaction you get from doing these things is worth every bit of effort. It is also worth more than any amount of money you might make along the way.

Raising chickens isn't difficult. Like anything else, it will turn out best when you give it your best. Or like your mom always said, "Anything worth doing is worth doing right."

If you've enjoyed reading this book, subscribe* to my mailing list for exclusive content and sneak peaks of my future books.

Visit the link below:

http://eepurl.com/glvBjj

OR

Use the QR Code:

(*Must be 13 years or older to subscribe)

www.ingramcontent.com/pod-product-compliance
Lightning Source LLC
Chambersburg PA
CBHW030017190526
45157CB00016B/3083